WISP Explained

Dr. Kalai Kalaichelvan
Lawrence Harte

DiscoverNet Publishing
1000 North Main Street, Suite 102
Fuquay Varina, NC 27526 USA
Telephone: +1.919.301.0109
email: info@DiscoverNet.com
web: www.DiscoverNet.com

International Standard Book Number: 9781932813586

i

About the Authors

Dr. Kalai Kalaichevan

Dr. Kalaichelvan is a wireless technology business expert with over 30 years of experience as an executive in the communications industry. Since founding EION in 2001, Dr. Kalai Kalaichelvan has grown the business to become a world leader in broadband wireless networking solutions and a pioneer in WiMAX, LTE, CBRS and 5G product development. Kalai's knowledge of the industry has led him to speaking engagements at the World Bank and the title Innovator of the Year from the National Research Council and the CATA Alliance.

Mr. Lawrence Harte

Lawrence Harte is a wireless communication and media expert. Mr. Harte is the founder and senior editor of Wireless Internet Service Provider - WISP Magazine - WISPMagazine.com/magazine. He has designed, installed and tested commercial and military RF communication equipment and systems ranging from voice radio communication systems to aircraft monitoring and analysis modules. Mr. Harte has written 121 books as of 2021 (30+ on wireless) and is the senior editor of Wireless Dictionary (wirelessdictionary.com). He was a voting member of the TIA Industry standards committees where he reviewed and created digital mobile radio industry specifications. Mr. Harte has been a speaker at 110+ wireless industry events. He created and presented the 3 hour New RF Technologies course at IWCE conferences for 10+ years. Mr. Harte is the inventor of multiple patents on mobile system radio signaling.

Expert Contributors

Jimmy Schaeffler

Jimmy Schaeffler is chairman/CSO of The Carmel Group, a telecommunications, computer, and media industries business analysis and expert consultant. Under the umbrella of The Carmel Group, Mr. Schaeffler has helped wireless communication and media companies to plan, value, and optimize their systems, since 1995. He worked with and interviewed hundreds of wireless Internet service providers to create the Wireless Internet Service Provider Associations (WISPA) 2021 "Lift-off" report, which covers key benchmarks and trends in the Fixed Wireless and Hybrid Fiber Wireless Broadband industry.

Cynthia Rawlings

Cynthia Rawlings is marketing director and promotions manager for multiple Wireless Internet Service Providers in Canada and contributing author for WISP Explained book and course. She is a guest host for WISP Podcast. Ms. Rawlings oversees brand development and management, marketing messaging and media, creates and coordinates promotion campaigns, sets up and manages over 60 media communication channels, works with media agencies and is responsible for ad campaigns and budgets.

Acknowledgements

Many smart people have helped to create this book. Some of them gave substantial amounts of time to share their experience, answer many questions, and invite us into their businesses and onto their production sets.

We thank the many wireless system owners and operators Aaron Remer from Tess, Jason Marsland with Velocity, John O'Farrell at Cochranetel, Nathan Stooke founder of WisperISP, Rick Harnish at Nexlink Internet, Sam Curtis from Atlink, Shaun Olsen leader of Cloudwyze, Sisso El-Hamamsy of Mage Networks, Sonia Winlad from Etheric Networks, Tracy Doaks with MCNC and Zot Barazzotte at Clifton Communications.

Many of the experts at EION Wireless and PomeGran shared their knowledge, experience and very patiently helped to explain many of the complex WISP system technology, business and marketing options. Thanks to Ashraf Youssef, Hari Acharya, Harshul Garg, Isha Kritika, Kasturi Muthukumaran, Manoj Awatramani, Ovi Biris, Prash Nedumaran, Robert Klein and several other staff.

Special thanks to consultants and research analysis including Simon Murray from Digital TV Research and Gladston Gomez media broadcasting consultant.

Thanks to the many equipment manufacturers, system developers and support service providers including Alison Vreeswyk with Preseem, Berge Kaprelian of Channelvisions Beka Media, Brian Young from IP Pay, Cesar Ruiz at Learning Alliance, Claude Aiken with WISPA Association, Daniel White with Atheral, Dave Thomas at VISP.net, Dennis Burgess from Link Technologies, Eric Hill with Enterprise Wireless Alliance, Gustavo Neiva Medeiros leader of WISP TV, Jeff Neblett from ISPN, Jonathan Black from CanWISP, Ken Janc with Market Broadband, Mike Wendy from the WISPA Association, Nathalie Shamir with Divi Networks, Ric Johnson at CityGrid, Sasha Sigal leader of CTI Connect, Steele Bennett with Federated Wireless, Stephen Coran attorney at Lerman Senter and Yisha Amsterdamer genius at Telrad.

Table of Contents

Foreword

by Jimmy Schaeffler

As of October 2021, in the United States there are an estimated 2,800 WISP service providers with approximately 7 million subscribers that are generating approximately \$5 billion annually. That revenue number is expected to grow to almost \$11 billion annually in the next four years, to 2025, as are the other numbers. For example, 13 million subs are projected for the U.S. fixed wireless and hybrid fiber wireless industry during the same timeframe.

WISP Subscriber Growth

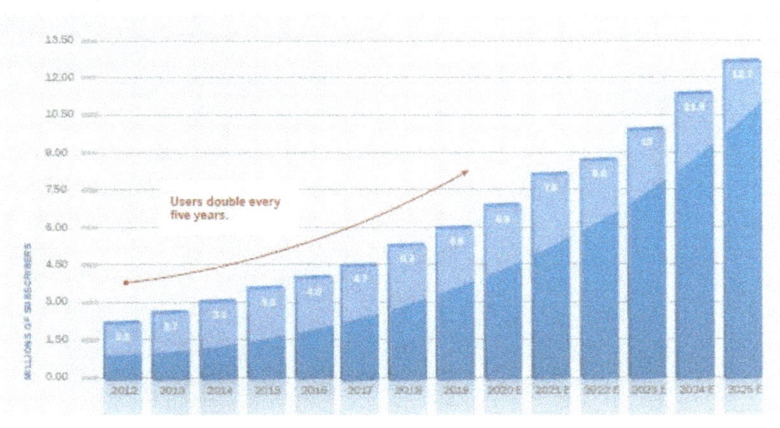

WISP Subscriber Growth USA Copyright, The Carmel Group, 2012-2022
All Rights Reserved

WISP Subscriber Revenues

The average revenue per unit from residential customers is approaching $60 per month, by the end of this year. In the next several years -- again, we take our projections out to 2025 -- thus we're estimating that ARPU will rise into the high $70s.

More importantly, the real bottom line for fixed wireless has become: because of its infrastructure; because of its affordability; because of the support it is starting to get from industry, government, and the financial communities right now, its operations and services will shatter the digital divide.

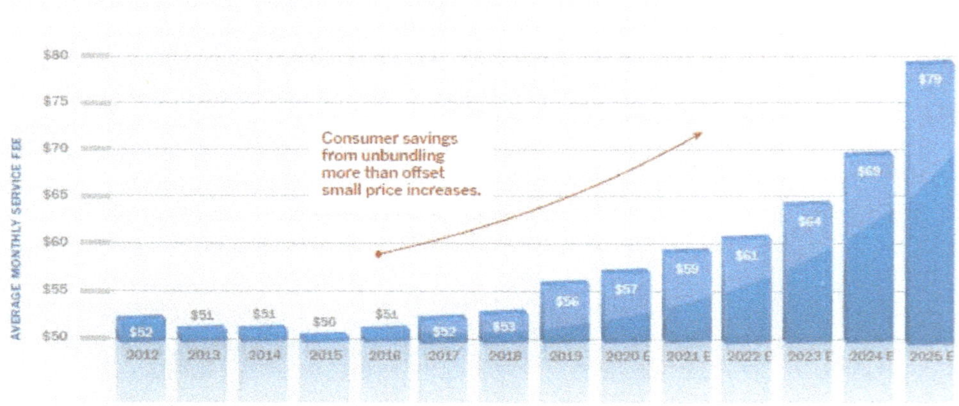

WISP Average Revenue per User USA, Copyright, The Carmel Group, 2012-2022
All Rights Reserved

WISP Marketplace Liftoff Report

To get key WISP industry market data, the WISPA Association has used The Carmel Group multiple times since 2012. We gather, analyze, and produce practical business reports for WISP system owners, investors, and regulators. The latest "2021 Liftoff" report is an industry broadband market report, covering the fiber and fixed wireless industry. It's a look at the broad industry with a lot of specific graphs involved, and really an effort to just tell the remarkably positive story of the United States WISP industry, circa 2021. The report includes case studies, market trends, forecasts, and growth drivers.

WISP Industry Data Sources

WISP industry business data is typically private and very hard to obtain. To gather, analyze and create reports for this data, The Carmel Group reached out to almost 800 survey takers embedded in as many U.S. based WISP hybrid fiber wireless operations, plus another 50 respondents from WISP vendor stakeholders, and conducted thirty personal interviews of roughly 90-minutees each.

You can download a copy of the one-of-a kind-in-the-world 33-page 2021 "Lift-off" report at no cost by visiting:

https://www.carmelgroup.com/2021-fixed-wireless-and-hybrid-fiber-wireless-report/

Chapter 1

What is a WISP?

A Wireless Internet Service Provider (WISP) is a company that connects user devices, such as computers, smart TVs, and security video cameras to the Internet by high-speed wireless connections. A WISP system owner purchases a high-speed link to the Internet, sets up radio access points (APs) which allow many users to connect to the Internet. The home devices typically first connect to a home data network (such as WiFi or a router) which are connected to a radio conversion device called a customer premises equipment (CPE) that links to a nearby WISP radio access point (AP).

This image shows that a WISP connects homes or businesses to the Internet using a network of access points, backhaul links and a high speed Internet connection point. Customers have a customer premises equipment (CPE) to link their home networks (convert to radio) to a nearby WISP system access point (AP). The AP converts signals from multiple customers into a format that can be transferred (backhauled) to a high speed Internet connection point. The WISP network management system (NMS) manages the customers, services and connection paths in the system.

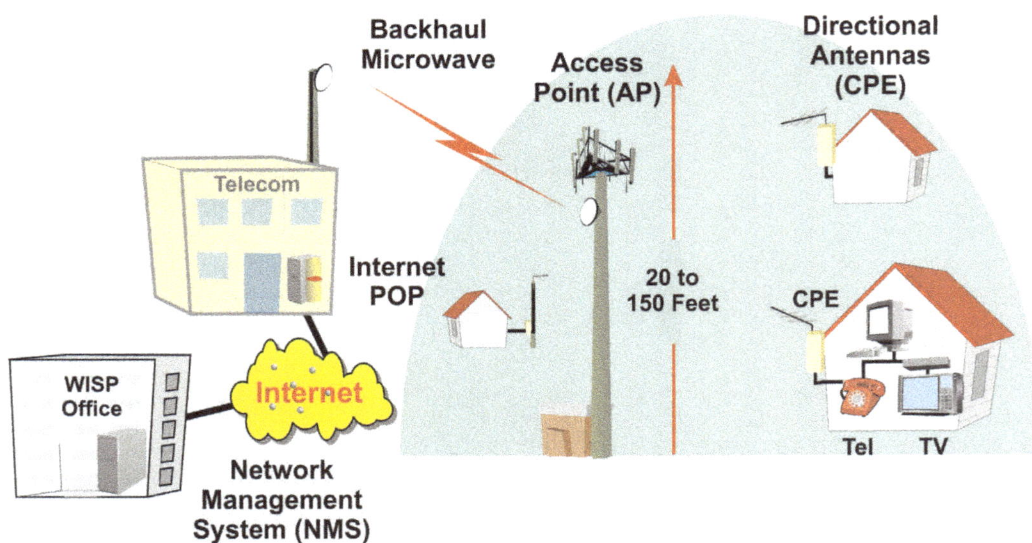

Figure 1.1, WISP System

WISP Business Benefits

To help motivate the building of WISP systems, many countries have provided free access to radio frequencies and offer government subsidies, especially to provide broadband Internet in rural areas. WISP systems can be rolled out in small areas - even small neighborhoods, have very low initial equipment cost per customer and tend to automatically operate without much help after they are setup.

As wireless technology improves, it is possible to provide higher speed connections with simple low cost system upgrades. WISP system operators control Internet access to their customers which can be used to promote and provide many types of cloud based services.

WISP Business Benefits

Continuous Revenue Streams

Free use of Shared Radio Frequencies

Lower Capital Cost per Customer

Government Grants and Loans

Flexible Build & Rollout

Automated Systems & Services

Bandwidth Upgradeable

Customer Access Relationships

Figure 1.2, WISP Business Benefits

Continuous Revenue Streams

WISP systems typically provide subscription services that are billed each month providing a reliable steady income. The cost to get customers and the disconnect rates are low, especially in rural areas where they have few (if any) other choices. The average revenue for WISP service in the United States was $59 in 2021 and is expected to increase to $79 in 2026 [Carmel Group, Liftoff Report 2021].

Free use of Shared Radio Frequencies

WISP systems can use shared access radio channels such as CBRS without having to buy wireless licenses or compete in spectrum auctions.

Lower Capital Cost per Customer

The cost to add WISP fixed wireless customers can be less than 10% of the cost of adding fiber customers [Carmel Group Liftoff Report 2021, page 18]. Each WISP access point can provide high speed Internet service to 20 to 50 customers with an average capital (system) investment cost of less than $500 per user [WISP operator interviews, WISPAPALOOZA 2021 show].

Government Grants and Loans

WISPs can receive multiple sources of free money (grants) and loans from governments and associations. These are often created to provide broadband Internet access to rural areas. WISP owners may get 80% or more of their initial cost without any repayment requirements.

Flexible Build & Rollout

WISP systems can start in small areas and gradually expand into other areas. It is ok for coverage for WISP systems to be targeted to specific areas. The cost of adding customers is a relatively low Initial Investment, typically 3x-10x less than fiber, coax or DSL.

Automated Systems & Services

WISP systems can use well standardized LTE technology which has self organizing network (SON) automatic network configuration capability - almost plug and play. Once the WISP system and customers are setup, the system tends to automatically operate - very low operational costs.

Bandwidth Upgradeable

Wireless technology continues to improve which means that data rates to fixed customers can continually increase without having to make major investments in new types of fiber, cables and network equipment.

Customer Access Relationships

WISP service providers are the connection access point for the customer. This allows WISP to promote and provide additional services such as phone, TV and cloud services. Cloud service partners may pay the WISP new customer activation incentives and offer revenue sharing options.

Broadband Requirements

WISP systems are developed to provide broadband Internet services. Broadband service requirements include data speeds for downlink uplink, services and applications and performance requirements.

Requirement	Description
Downlink Data	25 Mbps or greater – able to do multiple streaming services (NetFlix, etc)
Uplink Data	3 Mbps or greater – able to upload files or do video streaming
Services	Internet Access, Telephone, Monitoring, Streaming TV and other IP Cloud Services
Performance	Very short transfer delay (less than 1/10th of a second)

Figure 1.3, Customer Broadband Requirements

Downlink Data Rates

Downlink is the transfer of data from the system (nearby tower) to the receiving device (Internet connected devices). In general, WISP system downlink speeds can be faster than uplink speeds because of radio system technology. Broadband service with 25 Mbps downlink can provide for 2 to 3 simultaneous high definition (HD) video streams, file transfer and web browsing services.

Uplink Data Rates

Uplink is the transfer of data from the user (home or business devices) to a nearby WISP access point (AP). Broadband service with 3 Mbps uplink can provide video calls for 1-2 people and medium speed file uploading.

IP Service Applications

Service applications can work through the WISP system using standard Internet Protocol (IP) communication. These include telephone service (VoIP), monitoring (IoT and video), streaming TV and other cloud applications.

Service Performance

Fixed wireless systems can provide broadband signals with very short transfer delay (less than 1/10th of a second), consistent transmission quality (no large data gaps) and maintain high reliability (99.9%+) even in adverse conditions (rain, wind, etc).

WISP System Capabilities

Key WISP system capabilities include access point high data transmission rates, rapid response time, system reliability and low equipment & installation costs.

Capability	Description
Access Point Data Rates	Total data rates typically 200 Mbps+
Response Times	Less than 5 msec transmission.
Reliability	Able to achieve 99.9%+ uptimes
Equipment Costs	Less than $500 equipment cost per customer
Installation & Maintenance Costs	Fixed wireless installation cost can be 90% less than fiber or cable systems

Figure 1.4, WISP System Capabilities

System Data Rates

Each radio access point may have 200 Mbps total capacity for each channel which is shared by multiple users (typically 20-30 users). Because the average data rate used by each user is usually less than 10% of their available data speed, an access point with 200 Mbps capacity can provide 25 Mbps service to over 50+ users. As radio technology improves (new modulation and radio channel types), the available data rates will increase.

Response Times

Communication response time, the maximum amount of transmission delay (latency) can be less than 5 msec. Low latency delay is very important for live media (video, telephone and gaming).

System Reliability

WISP systems can often reach the 99.9% service uptime that wired ISPs promote. Fixed wireless WISP system failures can come from equipment lightning strikes and system failures (similar to other broadband systems) but WISP systems do not experience outages from cut cables or fibers.

Equipment Costs

WISP systems require radio equipment (CPE) to be installed at the customers location (home or business) which cost approximately $100 to $300 per unit. Each customer shares the cost of WISP access points (about 20 to 30 sharing each access point) which is approximately $50 to $100 shared cost per user.

Installation and Maintenance Costs

WISP system installation is fast (about 2-3 hours) and there is no maintenance cost for wireless transmission links. Broadband distribution cable and fiber installation costs can be $10,000 to $20,000 per mile and the cost to repair cable cuts can be $5,000 or more per splice [personal interviews with WISP system owners at WISPAPALOOZA 2021]..

WISP Competition

WISP provider key competitors include other WISPs, fiber, cable modems, mobile data, digital subscriber line (DSL), mobile data and satellite Internet.

WISP

WISP systems provide service to customers that are at fixed locations (fixed wireless broadband - FWB) within distances typically 5 to 10 miles of each radio access point. The maximum data rate is typically 200+ Mbps download and 50+ Mbps upload. It is possible for WISPs to setup wireless transmission with 1 Gbps connections for specific users. WISP wireless systems are continually evolving with more efficient modulation types and new radio frequencies so the data transmission rates and capacity of WISP systems will continually increase.

Fiber

Fiber systems are mostly available in urban areas due to the high cost of installation. Fiber systems can provide data services with 1+ Gbps download speed and 500+ Mbps upload speed. Some of the main challenges of fiber systems is the high cost for installation of fiber lines and equipment. Once fiber systems are setup, upgrading capacity and capabilities can be costly.

FIGURE 8: Comparative Economics of U.S. Internet Access Solutions

	FIBER	CABLE	SATELLITE	MOBILE	FIXED WIRELESS (1)
CAPEX/SUB RELATIVE TO FIXED WIRELESS (1)	9	5	6	3	1
YE 2020 BROADBAND SUBS IN MILLIONS (2)	15	76	2.5	315	6.9
EST. TYPICAL SUBSCRIBER SPEED DOWN (MBPS)	940	200	20	15	25
MAX SPEED DOWN (MBPS) (3)	2,000	2,000	50	200 +	1,000
UPGRADE COST	Moderate	High	Low/High	High/Modest	Low/Moderate
UPGRADE COMMENT	Only replace the endpoint; fiber remains the same.	Moderating with DOCSIS 3.1; less with linear TV.	Low incremental cost until the satellite dies; higher when satellite cost is included.	3G to 4G High; 4G to 5G Modest.	Requires modest incremental upgrades in CPE, towers, and networks.
AVERAGE REVENUE PER USER (ARPU) (4)	$65	$70	$90	$60	$57
CAPITAL OUTLAY PER SUBSCRIBER (INCLUDES PROVIDED CPE) (5)	$4,500	$2,200	$3,000	$1,300	$475
PAYBACK TIME IN MONTHS	69	31	32	22	8

1) This is a relative presentation comparing all of the technologies to fixed wireless, which is set to an index value of 10.
2) Subscriber numbers come from company filings, the CIA's "World Factbook," and estimates by The Carmel Group.
3) Mobile speeds can be higher during low traffic periods.
4) ARPU comes from a blend of advertised prices and company financial reports.
5) Capital Outlay figures come from company financial reports and estimates by The Carmel Group.

Figure 1.5, WISP Competitive Analysis - Copyright, The Carmel Group, 2021, 2022
All Rights Reserved

Cable Modem

Cable TV systems with cable modem services are available in most urban and some suburban areas but are rarely available in rural areas due to high cable installation cost. Cable modem services typically provide up to 200+ Mbps for each customer but could provide up to 4 Gbps for dedicated users. Cable systems have limited upchannel capacity which is shared by tens or hundreds of people.

Digital Subscriber Line (DSL)

Digital subscriber line (DSL) services are provided on twisted pairs of telephone wires. DSL services can be 100+ Mbps or more but speed decreases with distance - typically less than 10 Mbps at distances 5+ miles from the telephone switching center. This means that DSL data rates in rural areas can be 1 Mbps or lower.

Mobile Data (Cellular)

Mobile telephone systems can provide high speed Internet services - 10 Mbps-15 Mbps to each mobile device. Because the radio signals on mobile systems are shared in a cell site radio coverage area (not fixed and focused like WISPs), this limits the overall data capacity for each user. This means the cost to provide data transmission services on mobile systems is much higher (lower number of high bandwidth users per high cost radio tower) than fixed wireless systems. This is why mobile systems want to offload large data users (such as streaming TV) to WiFi and fixed wireless systems.

Satellite Internet

Satellite Internet systems use radio transmitters in space to relay signals. The download data rates for two-way satellite systems range from 1 Mbps to 150+ Mbps. Uplink data speeds for satellite systems can be 10% of the downlink rate. For example, a 25 Mbps satellite downlink data rate will have a 3 Mbps uplink data rate. Some satellite systems are geosynchronous (GEO) fixed locations and others use satellites in low Earth orbit (LEO) which continually move.

GEO satellites are located at 22,300 miles above the Earth which adds about ½ second delay to transmission (¼ second for transmission time + network transmission delays). LEO satellites are located at 340+ miles in the sky which adds less than 1/10th of a second delay.

WISP Funding

WISP systems can get funding from national and local grants, customer service fee subsidies, guaranteed loans, partner resources, customer pre-payments and other sources.

Funding Type	Description
Grants	National/federal and local government funds that are available to subsidize the installation and operation of broadband communication services
Customer Service Fee Subsidies	Government and business funds that are provided directly to consumers to reduce or eliminate their communication costs
Guaranteed Loans	Government, business or investor backed loans
Partner Resources	Vendors, dealers, or retailers who provide delayed payment terms or access to facilities or resources to help the WISP business
Customer Pre-Payments	Companies or people who pay connection request deposits advance service fees
Other Sources	Crowdfunding, community groups, municipalities, WISP friendly banks & other options

Figure 1.6, WISP Funding

Grants

Grants are money or resources from national/federal and local government funds that are made available to subsidize the development and operation of broadband communication services. The grants may be targeted to under-served rural areas and the types of broadband services (wireless or fiber).

Customer Service Fee Subsidies

Customer service fee subsidies are government and business payments that are directly received by the user to reduce or eliminate their communication costs. The WISP service provide may help customers to learn about and obtain these subsidies.

Guaranteed Loans

Loans guarantees from government, businesses or investors can help WISP operators get loans from banks and other funding sources.

Partner Resources

WISP partner companies (vendors, dealers, or retailers) may provide delayed payment terms, free or reduced costs for facilities use (buildings, towers) or resources to help the WISP business.

Customer Pre-Payments

WISP may offer discounts to customers who pre-pay for months or years of future services. WISPs may setup broadband service request registration systems that allow potential new customers to submit a deposit to confirm their interest in obtaining WISP service.

Other Funding Sources

Other WISP funding sources include crowdfunding, community groups, municipalities, banks that understand and loan to WISP businesses and other options..

Industry Standards

Some of the industry standards that WISP systems use include WiFi, WiMAX, 4G LTE and 5G NR. Industry standards define the radio access technology (RAT), features and may include some of the back end network requirements and capabilities.

Funding Type	Description
WiFi	Unlicensed wireless local area networks
WiMAX	Licensed alternative to wired DSL and Cable Modem
4G LTE	Broadband mobile packet data radio technology
5G NR	Multi-band, multi-technology platform that is 10x higher capacity than 4G

Figure 1.7, WISP Industry Standards

WiFi

WiFi is a group of wireless industry standards (802.11 series) that are typically used to allow devices to transfer data through wireless local area networks (WLANs) or directly to other devices (e.g. printers or media devices) with transfer speeds up to 10 Gbps. WiFi standards allow for the automatic discovery of devices and dynamic sharing of radio signals.

The typical maximum distances for WiFi systems is about 200 meters, it is possible to extend the range to over 5 km with directional antennas. While some WISP systems use or have used 802.11 WiFi systems, these systems and the frequencies they use (2.4 GHz and 5.7 GHz) tend to have many users which results in congestion and uncontrollable data rates.

WiMAX

Worldwide Interoperability for Microwave Access (WiMAX) is a set of wireless broadband communication standards launched in 2001 which was developed as an alternative to high speed wired connections (DSL and cable modem). The initial data speeds for WiMAX was 30-40 Mbps which has now increased to over 1 Gbps with technology improvements. Because of WiMAX limitations on uplink data speeds and limited management feature sets, WISP systems that use WiMAX tend to be shifting to 4G LTE systems.

4G Long Term Evolution (LTE)

Fourth generation long term evolution (LTE) is a broadband wireless industry standard that can provide up to 300 Mbps downlink and 150 Mbps uplink for each wide radio channel. 4G LTE is a version of the global third generation partnership project (3GPP) that has become the global standard for mobile communication systems.

The 4G standard is an evolved version of the 2G and 3G digital mobile telephone systems. Because there are millions of radio towers that use mobile cellular system radios, the technology and equipment has proven performance and the large economy of scale make the 4G radios and their systems relatively low in cost.

5G Mobile New Radio (NR)

Fifth generation new radio (NR) is an ultra broadband wireless radio system that can provide up to 20 Gbps data transfer speeds. 5G NR is a version of the global third generation partnership project (3GPP) that was developed to provide the types of services ranging from Internet of Things (IoT) to multi-gigabit mobile broadband communications.

5G NR is 10x+ more efficient than 4G LTE radio systems and this will allow each access point to provide service to hundreds of simultaneous users. Unfortunately, 5G NR wireless is not backward compatible with 4G LTE systems.

Chapter 2

WISP Services

WISP companies can provide Internet access, phone, home tech support, monitoring, TV and cloud based applications.

WISP service revenues can come from monthly service fees, activation commissions (new subscriber bonus), service promotion incentives (such as pre-loading a TV app in an access device), revenue sharing and other options.

This image shows a sample WISP service plan which includes Internet access, telephone service, monitoring service, television service and business applications.

Internet Access

WISPs can provide broadband Internet access services with multiple data rates (access packages). Internet access packages and rates can dramatically vary based on geographic region and competition.

In regions that have very limited alternative broadband Internet options may charge their customers $300+ or more per month for 5 Mbps access or less. Typical WISP monthly internet service fees range from about $20 per month for basic service (1-5 Mbps) to over $100 per month for 50 Mbps - 100 Mbps.

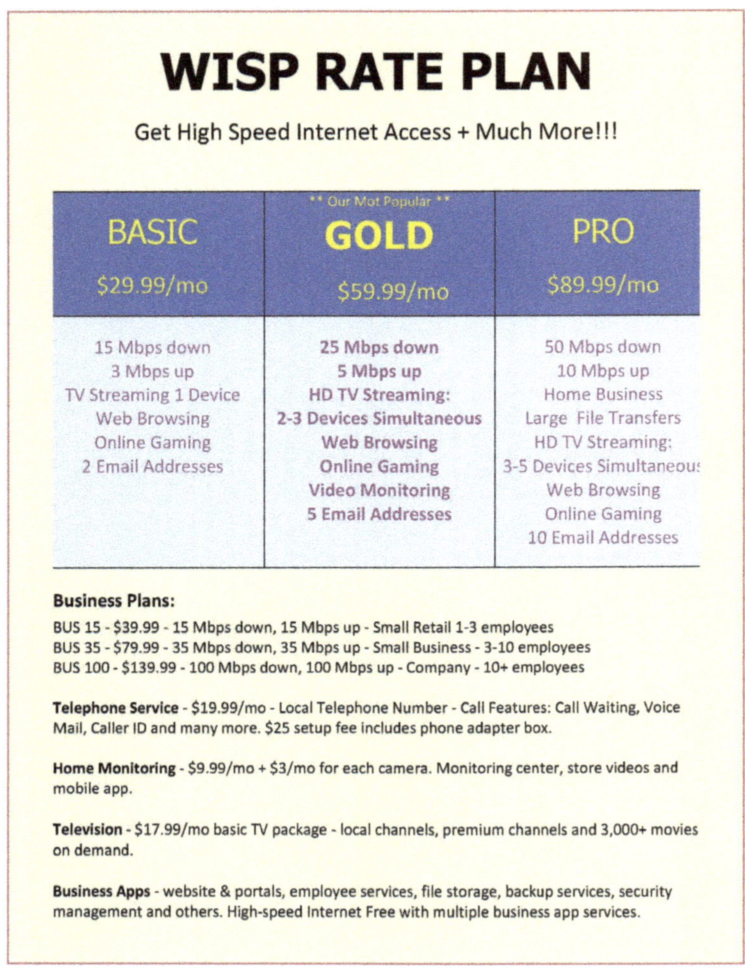

Figure 2.1, WISP Service Rate Plan

Data Rates

WISP Internet download data rates typically range from 5 Mbps to 100 Mbps. Upload data rates are commonly set at 20% to 30% of download rates. For example, an WISP Internet package that offers 25 Mbps download speed tends to offer 5 Mbps upload speed. For commercial/business accounts, the monthly service fees are 2x-3x higher than residential rates with the primary difference being higher upload data rates.

Service Packages

Internet access service packages may be described by identifying the typical types of user applications such as number of simultaneous streaming TV channels, file transfer size and other use types.

Monthly Data Caps

WISP may setup a maximum amount of data transferred each month for customers. High data usage costs WISPs more because they must pay Internet exchange point (IXP) access fees to connect their WISP system to the Internet. These costs typically range from $1,000 to $6,000 per month (co-location lease, equipment lease). If the WISP has 1,000 customers and has an IXP connection cost of $1,500 per month, it costs $15 per month for each subscriber. Offering higher data rates enables customers to consume a lot more media (such as streaming TV channels) which results in higher Internet network connection costs for the WISP.

Telephone

WISP service providers can offer telephone services with advanced calling features. Telephone service rates range from $10 to $25 per line and some WISPs charge additional setup and connection fees. A telephone digital media adapter (DMA) is required to convert the data connection to a standard telephone line.

Telephone Number

A WISP phone customer is provided with a telephone number that they can transfer from or to other telephone service providers. The WISP has to pay a monthly fee for each telephone number it gets and assigns to customers.

Telephone Adapter

A telephone digital media adapter (DMA) is required to convert the WISP voice over internet protocol (VoIP) into standard telephone signals. These adapters are small and cost about $20-$25 wholesale. Some WISPs give the phone adapters to customers at no cost.

Emergency Calling Capability

When providing telephone services, the WISP may be required to ensure it has emergency calling capability (with location identification) and backup batteries for power outages.

VoIP Phone Service Platforms

To provide telephone services, WISPs can become a reseller for a voice over Internet protocol (VoIP) cloud telephone service provider. The VoIP phone service provider may share 50% to 60% of the collected phone service fees with the WISP. The telephone services may be provided as a while label phone service allowing it to be branded with the WISP providers name.

Monitoring Services

WISPs can provide home security and video surveillance monitoring services. Home video monitoring service rates range from $10 to $35 per month [safety.com/home-security-cost/]. When a customer adds a security monitoring service, they receive a 2% to 10% reduction on their home and business insurance costs [policygenius.com/homeowners-insurance/how-much-can-you-save-on-home-insurance-with-a-security-system/].

Video Monitoring Service Platforms

Setting up and staffing a video service monitoring center can be complicated and expensive. WISPs may become a reseller for a cloud IP video monitoring service provider. The provider may share 50% or more of the collected monitoring fees. While label surveillance service providers brand the phone service with the WISP providers name.

WISP Tip - Avoid calling monitoring services as "surveillance services" because it can imply a responsibility for losses due to theft or may require government trade licenses for system installation.

Home and Business IT Support Services

WISP systems can provide home network setup, monitoring and IT support services. Fees for WISP customer home IT support can range from $7/mo to $27/mo or more. If customers do not use tech support services, the monthly service fees become all profit! Tech support services can be outsourced on a per call basis.

IT Setup & Maintenance Service

WISP operators can use contractors and 3rd party IT support companies to install home equipment, respond to calls and help requests and to provide answers to technical questions.

Network Monitoring Gateway

A key advantage of providing tech support services is the ability of the WISP to do network monitoring through the CPE equipment. This can help identify potential connection or media transfer problems that the customer has which is not caused by the WISP system.

WISP Tip - call WISP home IT support services instead of managed network services. Calling it managed home network services implies that the WISP may be responsible for problems or losses on the home network.

Television Services

WISP providers can offer broadcast and on demand TV services. Prices range from free (WISP earns from advertising revenues) to $99+ per month for high value streaming TV sports packages. WISPs can earn revenues from affiliate app install referrals (new account setup bonuses), by becoming a reseller for streaming TV providers (revenue share) or by creating TV apps that are branded with the WISP name (white label streaming TV apps).

Video Streaming

When WISP customers watch streaming TV, this uses a lot of data bandwidth. In addition to increasing the WISP Internet data connection costs, if many customers stream TV, this can require upgrades to the WISP data network - requiring more microwave links, increased packet switching capacity and other costs.

Broadcast TV Partnering

WISP companies can partner with broadcast TV providers to reduce the amount of TV streaming data use. WISPs can setup distribution relationships with satellite TV companies or local TV broadcasters. WISP customers watch most of their TV via the satellite connection and the WISP receives a portion of the satellite TV service fees.

Television Adaptor

WISPs can provide TV digital media adapters (such as Roku boxes) which allow the WISP company to manage the TV apps the customer uses. Companies that have apps which want to reach new consumers may pay the WISP an activation commission (bonus) for new customer accounts and/or provide a service revenue sharing commission.

Local TV Content

WISPs can support the creation of local TV content such as high school sports, community groups and other content. In addition to this content being low cost (or free) for the WISP TV system, local TV projects can provide a high level of awareness and have a very positive influence on the community.

Cloud Business Applications

WISPs can setup and provide access to applications such as database management, file storage, backup services, CRM, ERM, SCM and others. These can be provided directly by the WISP or through cloud service provider (CSP) services. Business application services can range from $10 per month for single apps to hundreds of dollars per month for bundled service packages.

In addition to generating additional high profit revenue for the WISP, once a customer has subscribed to business applications, they become highly sticky customers. If they leave to go to another service provider, they will have to setup new apps and move the data from their apps to the new platforms.

Bundled Internet Services

Because cloud business application revenue can be very high, some WISPs provide high-speed Internet connections for free when customers subscribe to their business app services.

Cloud Service Reseller

WISPs can sell cloud services by using staff, independent sales agents or by referring customers directly to CSPs. An easy way for a WISP to provide business cloud services is to become a CSP reseller. This means that the cloud services are setup and supported by the CSP and the WISP receives a percentage of the recurring revenue.

22

WISP Tip - identify cloud applications as IT or business access services. Calling cloud services as managed applications may imply responsibility for lost data when services or applications do not work correctly.

Chapter 3

WISP Regulations

WISP regulations include managing access to radio spectrum, Internet connectivity, grants & resources, national and local regulatory requirements and taxes & fees.

Funding Type	Description
Radio Spectrum Access	The use of licensed (exclusive), lightly licensed or unlicensed (shared & free) channels
Internet Infrastructure Connectivity	Direct high-speed Internet connections
Grants and Resources	Subsidies & allowances for underserved areas
National and Local Regulations	Telecommunication and information service rules and requirements
Taxes and Franchise Fees	Business taxes, sales taxes, franchise fees & other taxes agencies can add

Figure 3.1, WISP Regulations

Radio Spectrum Access

Communication agencies (such as the FCC, DOC, etc) define frequency bands with exclusive licensed) or shared (unlicensed) use and the services that can be offered through them. Unlicensed channels may self manage channel sharing (such as WiFi) or they may dynamically detect and manage interference between users (such as CBRS).

Providing WISPs with access to additional radio spectrum frequency bands is important so they can provide more broadband access for the new services such as streaming TV, distance learning and other applications. Associations such as the Wireless Internet Service Providers Association (WISPA) help the government agencies such as the FCC to understand WISPs growing needs for radio spectrum and ways to provide fair and reasonable cost radio access.

Licensed Frequency Bands

Licensed frequency bands provide authorized users with exclusive access. Governments may provide frequency band licenses for specific types of use (such as wireless Internet, public safety, utilities). They may also sell licenses at auctions. License rules can change and some uses can be eliminated.

Lightly Licensed Frequency Bands

Lightly licensed spectrum is an authorization for a company to use radio spectrum (frequency bands) provided when they pay a small fee which authorizes them to provide services according to the requirements of the license. This may be for exclusive use or for shared use.

Unlicensed Frequency Bands

Unlicensed frequency bands allow users to share frequency by following transmission, power level and interference control rules. Unlicensed frequency bands such as 2.4 GHz and 5.7 GHz can have many users which reduces the capacity to each user.

Channel Sharing Systems - CBRS and Others

Channel sharing systems allow multiple authorized users to share a frequency band by detecting and adapting their use of radio channels. These may be unlicensed channels such as the citizens band radio service (CBRS) available in the United States. To use the CBRS frequency band, companies must connect to a database which monitors and assigns access to CBRS radio channels.

Internet Infrastructure Connectivity

Access to WISP System High-Speed Internet Connection Points with Fast Access to Websites and Services. Regulations require Internet companies to provide fair direct access to Internet networks.

Point of Presence (PoP) Internet Connection Points

Point of presence (PoP) Internet connection points are locations which provide a high-speed connection to the Internet. These can be at a local telephone, cable TV company or at a data center with access to multiple network service providers. WISPs need to get (rent) a location inside the facility where they can install data gateways that connect their WISP system to a dedicated Internet connection point.

Dedicated Internet Access (DIA) Connection Types

Dedicated Internet access (DIA) provides a constant dedicated speed connection in both directions. Not all DIA connections are the same. DIA connection points with local or regional systems (local telephone or cable TV companies) may have multiple gateways in their networks located between your WISP DIA connection and Internet Exchange Point (XP) network locations (the actual Internet). These gateway systems may slow down connections to key platforms such as Facebook, NetFlix and others. If possible, get a DIA connection directly to an IXP.

Grants and Resources

Grants and resources are awards or gifts that may be provided by governments or organizations to achieve a consumer or business goal. Government grants for WISPs have requirements such as broadband services, geographic areas and types of systems. The requirements for getting grants can be complicated and costly.

The rules and requirements for grants are setup by agencies which may not have a good understanding of technology or practical implementation requirements. It may be necessary to work with government agencies and organizations in states or cities to adapt their grant requirements to make it possible to achieve the grant objectives. For example, some states in the United States provide grants only for fiber systems. This can be impractical in rural areas where fiber is costly or almost impossible to install.

WISP National & Local Regulations

WISPs can provide many types of applications and services which are regulated by government communication (e.g. FCC, DOC), commerce (e.g. FTC) and other national and local government agencies.

Government communication regulatory agencies define and regulate communication services provided by wired and wireless systems. This includes the types of services, authorizations and restrictions. Regulations include providing licenses to TV broadcasters and other services. Even though WISPs may be able to provide TV services, they may require licenses to provide broadcast television services.

Government commerce regulatory agencies define rules for fair trade to protect consumers and business rights. This can define and restrict what types of business WISPs are able to provide and how they may be protected from competitive antitrust and restraint of trade activities. As competition increases, larger communication companies may perform unfair practices to disrupt WISP services. Regulations also exist on data privacy such as government identification numbers (e.g. social security number), address, credit card & bank account numbers and other personal data.

WISP Taxes and Fees

National and local governments setup business taxes, product and service sales taxes and municipalities may have franchise fees. While WISPs may be currently exempt from telecom and other communication taxes, as WISPs become larger, these taxes and fees may also be imposed in the future.

Broadband Connectivity Fees

Broadband connectivity fees (e.g. Universal Connectivity Charges) may be added to invoices to help fund the development of broadband Internet in rural and underserved areas.

Emergency Services Fees

Telephone services may require the collection of emergency services fees (e.g. 911). This is to help support public safety access point (PSAP) communication services.

Franchise Fees

Municipalities that provide access to telephone poles, conduits or facilities for communication service companies for installation of cables may require the payment of a franchise fee. This may be a percentage of revenues earned by the communication company.

Other Taxes and Fees

Other taxes and fees may include state and county telecom surcharges, communication services sales tax and a mix of other state and local taxes and fees.

WISP Explained

Chapter 4

WISP Systems

WISP systems provide broadband Internet services using fixed wireless broadband (FWB) connections. The use of fixed locations instead of mobile service allows for focused and dedicated signals increasing bandwidth, capacity and reliability. WISP systems provide reliable high-speed data transmission rates with very low transmission delays (low latency) which can continually monitor and dynamically manage the bandwidth provided to each user.

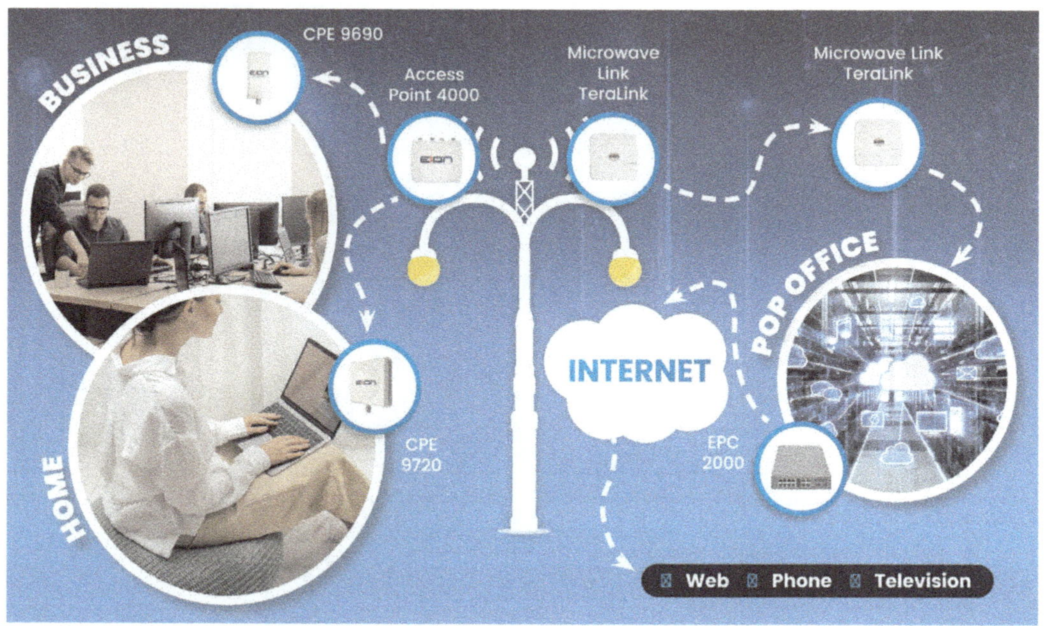

Figure 4.1, WISP System
Source: EION Wireless

Customer Premises Equipment (CPE)

WISP systems provide Internet services to customers through customer premises equipment (CPE) radio transceivers that are installed at the user's home or business building. These can be installed on walls, roofs or other locations that can be focused to a nearby radio access point (AP). Typically CPE equipment must have an unobstructed line of sight (LOS) to the AP. Installation usually requires a trained professional.

Figure 4.2, WISP Customer Premises Equipment (CPE)
Source: EION Wireless

The CPE usually includes a radio transmitter, data switching router and data connection that connects to the user's computer or their local area network (LAN). Power can be supplied to the CPE equipment on the same data line using power over Ethernet (PoE) which uses a small power supply in the house which puts a voltage on some of the wires in the data cable.

Access Points (APs)

WISP networks use radio access points (AP) to connect multiple users to the WISP system. Each AP provides service to tens or possibly hundreds of user CPE devices. An AP may be also called a cell site, radio port or access node.

Figure 4.3, WISP Access Point (AP)
Source: EION Wireless

Access Point Locations

APs are usually installed at a high location within 3 to 5 miles of users' CPE equipment. This may be on a light pole, top of a building or on an antenna tower. The WISP finds radio site locations (site scouting) and negotiates a lease agreement that allows for the installation and maintenance of the AP.

Access Point Power

APs require continuous power and typically have access to backup power that may run for several hours or days. APs may contain external or integrated antennas.

Backhaul Links

APs are connected to the WISP system's high-speed Internet connection point by backhaul links. These are typically direct line of sight microwave links but can also be fiber, coax or data lines. Because microwave links use highly focused radio beams, they usually don't interfere or get interference from other devices. This allows backhaul microwave links to use unlicensed frequency bands.

Figure 4.4, WISP Backhaul Microwave Link
Source: EION Wireless

Each microwave link can typically transfer 1-10 Gbps. If more data capacity (bandwidth) is needed, microwave systems can be replaced with higher speed equipment or additional microwave connections can be installed.

Microwave links require a line of sight (LOS) path between them. Because they use highly focused antennas (high gain), they can connect to distances over 10 km. They can use unlicensed frequencies because of their highly targeted radio beams and do not interfere with other devices or systems.

Radio Network Management

WISP radio systems can use multiple types of systems and channels. Many WISPs use the same long term evolution (LTE) system equipment that are used in 4G cellular systems. LTE is a proven standardized technology and the equipment is well understood and relatively low cost. The use of LTE equipment for WISP systems allows for mixing equipment from multiple vendor types - interoperability.

Figure 4.5, WISP Network Management - Evolved Packet Core (EPC)
Source: EION Wireless

WISP radio systems that use 4G LTE equipment have an evolved packet core (EPC) switching system which sets up and coordinates frequencies and manages connection paths through the system. The EPC oversees signaling on the control channels to use channel quality measurements to manage radio channels and connections through the system.

LTE WISP systems use a self organizing network (SON) to automatically select, configure and run the CPEs, APs and other network equipment. Epcs are a high-speed packet switching system that route connections through the network on the best available paths. If links become congested or some equipment fails, the epc will automatically reroute the connections providing for automatic and highly reliable operation.

PoP Internet Connection

A WISP system is connected to the Internet through a point of presence (PoP) high-speed Internet connection point. This is usually a fiber line connection that is located at a telephone system switching center or network providers building. The WISP system uses a gateway (converter) device between the WISP system and the Internet. The PoP owner charges are lease fee and additional monthly fees based on the amount of data transferred through the PoP. This means that WISP systems that consume a large amount of data, such as subscribers watching lots of streaming TV, will have higher costs.

Network Management System (NMS)

WISP systems use a network management system (NMS) software or a cloud service to setup, control, monitor, and manage the services provided by the WISP network. NMS include application program interfaces (APIs) that allow adding and interworking with other services including customer relationship management (CRM), billing and security.

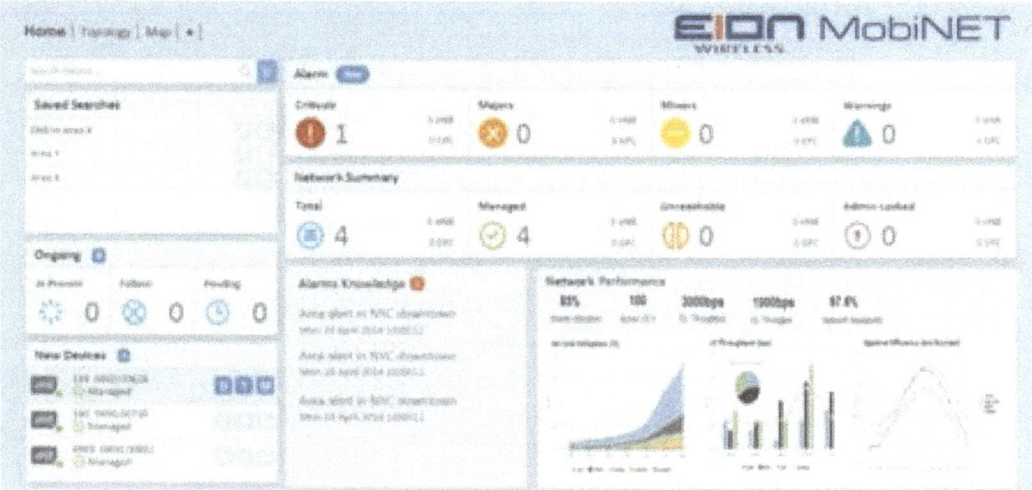

Figure 4.6, WISP Network Management System (NMS)
Source: EION Wireless

The WISP uses the NMS to connect with and setup network equipment and services - provisioning. The NMS is setup with the service packages (service ID, bandwidth, etc) and manages customers (customer ID, service packages, etc).

Equipment Monitoring & Control

The NMS may need to control different types of equipment from multiple manufacturers. To allow interoperability with different manufacturers, NMS uses standardized communication and control process simple network management protocol (SNMP). SNMP can communicate management information between the network management stations (NMS) and the agents (ex. routers, switches, network devices) in the network elements. By conforming to SNMP protocol, equipment assemblies that are produced by different manufacturers can be managed by a single program.

NMS Co-Existence

WISP systems may be added to an existing Internet service provider (ISP) system that already has a NMS. The existing NMS may not have the capabilities to setup and manage the WISP network or to setup and manage customer CPE equipment. This means the existing ISP NMS must work with the new WISP NMS.

IP Applications

WISP systems can offer multiple services including phone, security, streaming TV and many other types of cloud (Internet based) applications. WISP systems owners may purchase software and/or partner with IP application service providers to offer these services.

Adding IP applications to WISP systems can range from installing apps on customer devices (such as a streaming TV app on smart TVs), to installing fully integrated applications such as telephone services with advanced features. Applications may need to be configured for quality of service (QoS), bandwidth and other capabilities. It can be helpful to add applications from companies that have already been setup and tested with the equipment that is used by the WISP system.

Some applications may require additional subscriber equipment or to configure equipment the customer may already own. For example, a WISP that provides telephone service needs to communicate through a telephone adapter. Some devices may already have phone adapters or the WISP may need to provide a telephone digital media adapter (DMA).

Chapter 5

WISP Installation and Setup

WISP system installation and setup involves creating service orders, installing customer CPE, installing access points and service provisioning.

WISP Service Orders

WISP service orders are documents or files that identify customer service requirements, location review, equipment that will be used and their installation schedule.

Customer Service Requirements

Customer service requirements include the location (address) where the services will be installed, what services they want (services list). A customer service order typically includes service ID codes that will be used to setup (provision) the services ordered by the customer.

Location Review

One of the first steps in setting up an installation service order is verifying that the location is in the WISP system radio coverage area and that the building has a line of sight (LOS) access to a WISP access point.

Equipment Acquisition

The services ordered and location of the customer (building type) determine what customer premises equipment (CPE) can be used (radio type, antenna, data connection). If the equipment is in inventory, it can be obtained by or provided to the installer. If it is not available, the equipment may be ordered which can delay the installation schedule.

WISP Tip - keep several CPE radios on installation trucks and store some CPE equipment units at multiple locations (depots) in the WISP service geographic areas (installer homes, storage units, etc).

Installation Schedule

Customer CPE installation dates and times can be reviewed to determine when installers are available and near the customer location. Scheduling can start simple with Google Calendar and progress to a field service scheduling program or cloud management service.

WISP Tip - if you allow installers to park their trucks at home, use the installers home location for the drive to the first installation location.

WISP Service Location Review

When a customer requests service at their location, a service location review needs to be done to determine if they have access to a nearby WISP access point.

WISP systems usually provide services to limited geographic areas. This means that not all homes or businesses will have access to WISP radio signals. In addition to being inside a radio coverage area, customer locations need to have line of sight (LOS) unobstructed paths to the access point.

Service Location Access Verification Tools

WISP service location access verification software tools that display the customer's area on a map along with the WISP radio tower locations. The customer or sales agent enters the address where the CPE receiving equipment can be installed.

Available WISP Access Points

After the customer's location is entered into the location verification tool, it will show access points that are within a distance that can be used for a connection (there may be more than 1).

Access Point Line of Sight (LOS)

In addition to needing to be near a WISP radio access point, the customer needs line of sight (LOS) review. LOS is an unobstructed path between the customer's WISP CPE antenna and the WISP access point antenna. To help determine LOS path, location verification tools may include ability to overlay Google maps and government information service (GIS) terrain data. If a LOS can't be clearly identified from the visual tools, an onsite evaluation may be needed.

WISP Access Point (AP) Locations

WISP access point locations are selected to provide service to geographic areas. A typical WISP access point can provide radio coverage to a distance of 3 to 5 miles. WISP access points setup involves location scouting, electric utilities and maintenance access.

Location Scouting

Finding locations that are good candidates for WISP radio access points is called location scouting. WISP access points can provide radio coverage to neighborhoods, commercial districts or other areas. Location scouting can search for existing radio towers, buildings or structures which are good candidates for antenna location and height.

Utilities

WISP access points locations need access to electrical power for operation. The location owner needs to be willing to provide electrical service for your access point or authorize the installation of electrical service. For remote locations, solar power may be an option for electrical power.

Location Servicing Access

WISP access points need to have access for maintenance and repair. Each WISP access needs instructions on how WISP installation and maintenance staff can get access to the access point. Locations may require rooftop access which may not be easily available in the evening hours when the building is closed.

Access Point Site Leasing

Installing access points usually requires site leasing agreements. WISP access point site lease agreements typically provide monthly or annual lease fees and utility costs in return for the authorization to install, run and maintain the access point.

Site Location Leasing Fees

Access point location leasing fees can be $50 to $200 per month in rural or small town areas. In urban areas at popular locations (existing towers), site location leasing fees can be $1,000+.

Utility Costs

Access point site lease agreements may include access to electric service as part of the lease fee or it may require the WISP operator to setup or pay an extra fee for access to power. Access points typically don't use much power - about 300-500 Watts per radio.

Installation and Access Requirements

The lease agreement should include authorizations for access to perform testing and maintenance along with contact information needed to get access to locked areas.

Barter Agreements

WISP access point locations may be obtained at reduced or free cost site location leasing in return for Internet service or other types of value. In rural areas, land or building owners may allow antennas on their barns, grain silos or other structures in return for Internet access service.

WISP system installation and setup involves creating service orders, installing customer CPE, installing access points and service provisioning.

WISP Systems Integrators

WISP companies can get help from systems integrators to design, install, configure, integrate and setup operations for wireless Internet service provider (WISP) systems. What surprises some WISP operators, systems integrators get reseller discounts from suppliers that can reduce or even eliminate the added costs of using system integrator services. WISP companies don't have access or can't afford to hire people with all the skills needed. Multiple equipment and services can be hard to connect and integrate. WISPs need to understand business and marketing for them to make a profit. Understanding key systems development needs can help WISP to choose which systems integrators can help them to select, build and optimize their WISP system.

Figure 5.1, WISP Systems Integrator Services

There are tens of thousands of systems integrators with experience in communication systems and technology. There are a very limited number of systems integrator companies that have successful experience with wireless internet service provider (WISP) and CBRS systems.

Some systems integrators offer a "turn-key" system setup where they can select, install and begin operations for a WISP system. After the system is setup, the WISP operator then takes over the operation of the system. System integrators may offer network operations center (NOC) monitoring services and support after the WISP system is setup.

Systems integrator service contracts tend to be lengthy and complex. There are typically objectives, schedules and potential penalties if deadlines are missed. WISP system owners must provide access and resources to the systems integrator companies and their teams. The systems integrator must get approval for designs, selected equipment and installation plans. The contract needs to allow flexibility for unexpected needs or obstacles.

Figure 5.2, WISP Systems Integrator Services

Design

WISP system design must consider the types of services, number of customers and geographic area terrain that will be covered. Providing Internet services by wireless has some unique challenges and advantages compared to wired DSL, cable modem or fiber alternatives.

The systems integrator must understand how and where the WISP system will be used. This starts by defining the types of wireless services and capabilities the WISP wants to offer. Most WISPs start by offering multiple Internet access service plans. WISPs that offer and support services such as streaming TV will need to setup systems with higher capacity. The potential customer market size should be reviewed to estimate the number of customers and the types of services they are likely to want.

A WISP can choose from unlicensed or licensed frequency bands. The systems integrator may request and renew frequency licenses with the FCC. Systems integrators can help to identify and get leases for site location on existing towers, buildings, structures and locations. The system design is created by wireless and electrical engineers, solution architects and field service technicians who use radio planning tools. Test transmitters and receivers may be used to verify coverage areas.

Installation

WISP companies need to understand how to select and install equipment on radio sites, homes and businesses. System integrators can recommend and may be authorized to directly purchase and install equipment from preferred or specific vendors. The installed equipment may need to be tested to ensure the required performance has been achieved.

Configure

After equipment is installed, it needs to be setup, configured and connected to the WISP network. The equipment needs to be connected to the network management systems (NMS) which can monitor and control equipment and services.

Integration

WISP networks contain multiple equipment units, software and platforms from different vendors that need to be linked and communicate with each other. The systems integrator may install links, configure APIs and setup communication between systems and services

Operations

WISP system operations include monitoring, provisioning and management of customer and network equipment. Operations typically include a network operations center (NOC) that provides a visual display of system equipment, content and reports. The NOC may be initially located at the systems integrator company and transferred to a WISP controlled NOC. Monitoring devices with alert messages are setup to identify when trouble levels or failures occur and trouble ticket systems are used to track troubleshooting activities. Systems integrators may create or assist in the creation of operations and maintenance procedures.

Training

Training is a continual process as WISPs continually add new services and applications and upgrade equipment. System integrators can instruct and assist WISP staff to install, configure and provision equipment and services. This can be a mix of online courses, training materials or on site exercises to help staff to become familiar with equipment and systems. Training and certification programs can be setup to ensure existing and new staff have the skills and experience necessary to run and maintain their WISP systems.

Chapter 6

WISP Marketing

WISP marketing is the communication of awareness and motivational messages to people who are likely to be interested in WISP services and products. Good WISP services marketing programs can be low cost, work fast and be simple to manage. WISP companies are often surprised to learn their least cost media channels can have the highest overall marketing cost. Effective WISP customer acquisition marketing campaigns such as door hangers, direct mail and referral programs can have subscriber acquisition cost (SAC) from $18 to $100.

Promoting and marketing WISP services tend to have high costs for advertising, media creation, marketing staff & contractor wages. While the small businesses administration recommends a budget of 7% to 8% for marketing, WISP companies tend to be run by technical people who have limited marketing experience and sometimes a dislike for promotion activities.

WISP Marketing Differences

WISP services may be the only high speed Internet option available to customers in certain areas. This means that marketing campaigns with effective messages and offers can have much higher conversion rates (% of customers who buy). Even when WISPs operate in areas with competing ISPs, WISP service packages can provide unique value to customers.

WISP Subscriber Acquisition Cost (SAC)

WISP subscriber acquisition cost (SAC) can range from below $10 for word of mouth marketing to over $400 per customer for untargeted marketing campaigns.

WISP Marketing Options

WISPs need effective marketing campaigns to get new customers with low customer acquisition costs. Traditional marketing and advertising campaigns are costly and tend not to work well for selling broadband Internet services. WISPs typically do not have a lot of marketing skills or resources. Campaigns that receive many rejections from people can dishearten or unnerve staff.

WISP Marketing Campaigns

WISP marketing campaigns are promotion projects that help potential customers to discover, engage and develop a value understanding on how WISP services can help them. WISP Marketing campaigns are important to get new customers, increase value awareness of existing customers and to measure the success of marketing activities.

What surprises WISP companies is that marketing is not hard, initial marketing campaigns typically fail and the best marketing comes from data you already have. Some of the most effective WISP Marketing campaigns include Direct Mail, Local Sponsorships, Door Hangers, Referral Programs, Email Marketing, Service Tables & Booths, Videos, Signage, Reputation Management, and Internet Training.

Funding Type	Description
Review Marketing	Helping happy customers to go to review sites (e.g. Yelp) and post review ratings & good comments.
Direct Mail Marketing	Sending cards, letters and materials by postal service to potential customers
Door Hangers	Hanging cards or brochures on doors and entry areas
Service Tables & Booths	Setup tables or booths at flea markets, local events & shopping areas which have subscriber signup promotions.
Referral Programs	Asking and rewarding customers for introductions to their friends & associates
Reputation Management	Monitoring, responding and managing reviews and media that mention your company or products.
Customer Training	Run Internet training courses to help people learn how to connect to and use the Internet.
Email Marketing	Sending value messages and offers to targeted email recipients
Local Sponsorships	Providing money or resources for local groups in return for branding and messaging options
Co-Marketing	Exchanging and sharing marketing activities with related company types
Targeted Advertising	Paying to insert media into channels that target qualified potential customers
WISP Signage	Place images and messages where existing and potential customers can see them.
Word of Mouth Marketing	Get people to share helpful and positive information about your company
Video Marketing	Create and use videos to attract and motivate customers.
Social Media	Creating, sharing and engaging with social media platforms such as Twitter, Facebook, Linkedin, Instagram, Youtube and other platforms

Figure 6.1, Key WISP Marketing Options

Review Marketing

Review marketing are processes that help people to post reviews or comment about your company on review websites such as Yelp, Google places and others. Companies that have very positive reviews (more reviews and many stars) tend to be shown more in search engine results. Ask happy customers for reviews.

WISP Tip - Create Review Business Cards with links to key review websites: Yelp, Google Maps, Facebook and others. Train your installers and staff to share the review business cards with happy customers and suggest some topics or keywords that the person can talk about when posting the review such as "reliable", "low price", "better Internet service" and other key words or phrases potential customers value and are searching for.

Caution, review websites may have restrictions on asking for or influencing (rewarding) for reviews. Check the terms or service for review websites before requesting or giving customers specific recommendations for reviews.

Direct Mail Marketing

Direct mail marketing is the sending of materials through the postal service to create awareness, positive thoughts or offers for your services or products.

WISPs can do direct marketing themselves in small batches. You can make your own lists by searching through directories, get (barter) from other local companies that already have maikling lists or purchase postal mail lists for a small fee.

When renting or bartering lists from other companies, a mailing house or letter shop may be used to process the mailing. In addition to performing the handling the mailing processes (envelope stuffing and addressing), a mailing house may keep the list renter from obtaining the list, preventing them from using the list multiple times.

The content contained in direct mail marketing campaigns may include offers that involve responding to the mailing literature. Response channels to direct mail may include web site forms, email messages or telephone numbers.

Direct mail marketing typically costs 50 cents to 1 dollar each for the mailing cost and media materials (letter, brochure) plus 20 to 50 cents per person for list building or rental. Direct mail purchase conversion rates can be about 1% to 2% for a good campaign which is about $50 to $100 customer acquisition cost.

Direct mail projects get new customers within 1-2 weeks. Continual use of direct marketing can build brand awareness and value perception if the direct marketing materials offer value messages that build on each other.

WISP Tip - to get almost all people to open postal mail from your company, send letters with hand written addresses and use a real postage stamp with either no return address or a return address that is personal and hand written. Almost all people will open this type of mail.

Door Hangers

Door hanger marketing is the placing of promotional brochures on the doors of homes or businesses.

Start by identifying neighborhoods and areas that are in need of good broadband Internet services - underserved areas. Door Hangers usually cost 20 to 50 cents for each house distribution (about 50 per hour) and and 10 to 20 cents print cost for the paper hangers. Important - there can be restrictions on leaving door hangers on private property. The acquisition cost for door hangers tends to be about $25 to $50 for well run campaigns.

Results from door hanger campaigns typically happen within 1 to 3 days after the distribution of door hanger materials. Conversion rates can be 2% to 4% if neighborhoods in need of broadband are identified and targeted.

Caution - there can be restrictions on leaving door hangers on private property. The acquisition cost for door hangers tends to be about $25 to $50 for well run campaigns. Check with community organizations or local regulators on restrictions on leaving brochures or door hangers.

Service Tables & Booths

WISP operators can setup tables or booths at flea markets, local events and shopping areas which can provide special subscriber signup promotions.

The cost to rent a spot for a table for a day or two at a location can be small. You should have some signage (banner), a special offer (special discount or gift) and basic literature. While it is helpful that people who run the booth have an outgoing personality, good attention getting signage promotional items can do great at motivating people to come to your booth.

Prepare some draft sales scripts (key talking points), practice and provide bonus incentives for the people who man the booth for getting subscriptions ($10+ each) and contacts ($3 each for contact information).

A majority of the new subscriptions are completed during the event and a single day can get 10 to 30 new customers.

Referral Programs

Referral programs reward customers for providing contact information of friends or associates who may buy your WISP services or products.

Referral programs usually cost $10 to $25 for referral bonus payments plus program management (tracking and paying) of about $2 per referral (5-10 minutes) and 10 to 20 cents per referral request message (promoting the referral program to customers by email or phone). Because the WISP already has a relationship with the customer, it may be ok to email or call WISP customers without restrictions. Referral programs tend to have a 1% conversion rate resulting in an average acquisition cost of $30 to $45 ($20 promotion + referral bonus).

Reputation Management

Reputation management involves monitoring, responding and managing reviews and media that mention your company or products.

It is important to monitor and manage reputation because unhappy people tend to post negative reviews and happy people tend to not post reviews. If you do nothing, it is likely that a higher percentage of reviews will be nega-tive.

Review ratings (stars) influence the ranking in search results and can strongly influence the company trust and value for potential customers. Key review websites include Google Business, Yelp, Angie's List and local business directories.

To increase the number of positive reviews, it can be helpful to ask happy customers to post reviews. When possible, ask the person to comment in their review about key topics such as "helpful", "good value", "quality work" and other searchable and motivational keywords. Be careful, review websites have terms of service that typically limit and they may prohibit companies from requesting or influencing reviews.

Reputation management typically takes weeks and months to build and maintain. If you don't do reputation management and you have bad reviews, it can take months to rebuild a good reputation.

Customer Training Programs

WISP companies can develop and run Internet training courses to help people learn how to connect to and use the Internet. These can be provided in person or online.

WISP customers don't buy Internet services. They purchase the things Internet connectivity can do for them - streaming TV services, gaming, school and work programs and other applications.. Helping people to understand which applications can help them and how to use them can be an important valuable for your customers.

Courses can be hosted by schools, community groups and events. These hosts provide a location for the courses and typically will promote them to their members and associates. Providing Internet courses is a great way for your company to be seen by many people with a very positive view.

It can take months to create, setup hosting relationships and run training courses. While there will be some subscriber growth during the announcement phase, most of the new subscribers come during or after the event. The awareness and value benefits of educating customers can be long lasting.

Email Marketing

Email marketing is sending messages to people on contact lists to get them to do things.

Email marketing involves getting access to lists, creating value messages and tracking and optimizing desired activities - conversions.

You can own, rent or barter access to email contact lists. When sending email messages to lists, it is important for the sender to have a relationship with the recipient. While you need to conform to email SPAM regulations that prohibit sending of unsolicited messages, sending mail to people who don't know you often goes into the recipients SPAM folder. Whoever owns the list should send the message.

Use an email service provider platform such as Mailchimp (simple) or others to help organize your contact lists, create messages and track delivery and engagement (click) information. It costs about 1 cent per content per month to manage your lists.

Create related value messages that build relationships and trust for your company. Connect these together to form campaigns over weeks or months. It can be Internet training sessions, sponsored events, guides or other valuable information. If you send promotional offers they should occur after you have sent some value focused email messages and developed trust.

Review email campaigns that other SUCCESSFUL WISPs have send. Learn what works for them. Try some of the things they are doing and see if it works in your area. Track your results (clicks, contacts, purchases), try new things and continually optimize your email marketing content and campaigns.

The results of an email marketing campaign are typically completed within 1 to 5 days after a message is sent to the audience.

Local Sponsorships

WISP operators can provide money or resources to community groups or people to include or send messages to their members. Sponsorships are a great way for companies to show their commitment to support the local community which builds trust and value.

WISP companies can sponsor events, groups, local sports teams and other organizations. Sponsors may include WISP logos or messages on apparel (sports shirts) on program media materials (signage) or create new media that include sponsorship messages or interviews.

You can ask the organization to provide multiple media options and activities. Announce your sponsorship to their connections, introduce you as a sponsor at meetings or events, add your company information to a sponsorship page on their website with links to your company and other options.

Local sponsorships can rapidly develop awareness if your local sponsors email or message their customers or connections. Local sponsorships can provide awareness for months or years when your brand logo or message is included in the organization's media.

Co-Marketing (Marketing Partnerships)

Co-marketing partnerships are the exchange of promotion activities between companies. WISPs can setup co-marketing activities with companies that have relationships with people who are good candidates for WISP services. Co-marketing programs provide access (reach) to people who work with and trust the partner. Co-marketing projects may be simple undocumented relationships (you send our message to your costumers and we will send your message to our customers) or they may be revenue sharing or activity based.

An example of a WISP co-marketing program can be point of sale (POS) WISP brochure displays at computer and electronics stores. The brochures can include a promotional offer with a discount (tracking) code. When the customer uses the tracking code, the co-marketing partner earns an activation bonus.

Targeted Advertising

Targeted advertising is paying for the insertion of media into communication channels with well defined audiences (e.g. Google PPC Ads, Facebook Audience Targeted).

Pay per click (PPC) advertising on search engines such as Google insert ads when people search for keywords or phrases. PPC advertising is a good tool to get people who are actively searching for products or services. PPC advertising programs typically charge a fee based on clicks - not conversions (actual sales). The average cost per click on Google in 2020 was $2.69 (for ads in Google results) and $0.63 (for ads in website partners - display ads) [https://thinkresultsmarketing.com/google-adwords-industry-benchmarks-for-2020/]. It can take tens or hundreds of clicks to get each sale.

Facebook and other social platform ads can be targeted to audiences with specific characteristics, interests and activities. Because the viewers are not actively searching for the topic, Facebook ads are interruption based. The ads must stand out and interrupt the viewer in a way that motivates changing their current activity. Facebook and other social media ads can be targeted to audiences with specific characteristics - location, interests, etc. If you have Facebook pages with followers, you can ask Facebook to display ads to people with similar profiles (look alike) audiences. If your Facebook page followers don't have an interest that matches your product or service (you got lots of Followers - for any reason), your ads will be shown to the wrong people - typically with bad results.

WISP Tip - talk to other WISPs and find out which advertising programs are working for them and adapt their campaigns and media materials to your area.

WISP Signage

Signage is the placing of images and messages where existing and potential customers can see them.

Signage typically has the lowest cost per impression - lots of people drive by road signs. However, signage typically does not generate calls or website visitors. Signage is very good at developing awareness and brand value.

WISP companies can put signage on buildings, billboards, vehicles, equipment, lawn signs and other places. You can put signage on trucks (wraps) and you can park trucks almost anywhere for periods of time - shopping centers, office buildings and other locations.

There may be zoning regulations and signage restrictions in communities. Even if you get a business to approve you putting a sign in their front area, it may not be allowed by the city or zoning authority.

Signage takes time to work but when setup, it can provide long term awareness recognition and brand value message value.

Word of Mouth Marketing (WOMMA)

Word of mouth marketing is the promotion of services or products by people who use or know your product or service. WISPs can amplify and increase WOMMA by encouraging happy customers to share, asking for names of other people and by providing references to neighbors and other customers.

Video Marketing

WISPs can create and use videos to attract and motivate customers.

What surprises many people is that a key value of video marketing is the Video Title. Search engines like Google list websites and videos as separate links in the search results. WISPs that use videos for marketing get a boost in their search results regardless if the videos are any good.

The types of videos for WISP marketing include how-to, entertainment, testimonials and others. Video production can be extremely simple. It can range from doing screen captures while accessing a WISP service to video recording a testimonial. Videos do not need to be highly professional. Very slick videos may not be trusted by the recipients.

Most viewers will watch your videos for a very short time. When possible, offer something of value to qualified viewers within the first 10 to 20 seconds. Add your WISP logo and web address to your videos in the bottom corner.

Video marketing can take weeks or months to provide benefits. It takes time to get discovered and indexed by search engines. However, if you create a good library of valuable videos, the benefits can last months or years.

Social Media

Social media marketing is the creating, sharing and engaging with social media platforms such as Twitter, Blogs, Facebook, Linkedin, Instagram, Youtube and other platforms.

You should setup 10+ social media profiles (Twitter, Facebook, Instagram, etc) with your company name. In addition to your customers expecting their Internet service provider to have social media pages, you should own and control social media profiles with your company name. Having multiple social media profiles will also result in your WISP company showing up more often in search engine results lists.

While owning and publishing on social media channels may be free, the cost of writing, publishing and responding can be high. Social media channels such as Twitter and Facebook are free to use but few people will see and respond to posts. Each post can take 30 to 60 minutes to create and publish. An additional 30 to 60 minutes may be spent on monitoring and responding to social media. This brings the cost to $50 per social media post (about 1 to 2 hours of staff time per media post). Unfortunately, most media posts are not seen and the response rate to social media posts tends to be less than ½%. Social media post responses are typically likes, shares or comments that do not directly lead to product or service sales.

WISP Tip - Setup separate Facebook pages for WISP company information and system status updates. This will allow potential new customers to focus on the services and benefits of the WISP system and not be distracted by service interruption messages or bad feedback.

Media Channels

WISP marketing media channels are any service or media platform that can share messages between WISP system promoters and audiences. Media channels may be used to send (push) messages or they can be discoverable (pull). WISP system owners, staff or marketing agencies can own or manage WISP marketing media channels that send or share information with customers and prospects.

Funding Type	Description
Website	WISP Business Website - Services & Business Information
Facebook Business Page	WISP Business and Community Activities
Linkedin Company Page	WISP Business and Business Activities
Instagram	WISP Business Activity Photos
Youtube	WISP Business Services & Helpful Videos
Twitter	WISP Business Announcements and Value Messages
Reddit	WISP System & Service Discussion Topics
Google Maps	WISP Business Description & Reviews
Yelp	WISP Business Description & Reviews

Figure 6.2, WISP Media Channel List

Many WISP marketing media channels have no cost for publishing and distribution (such as Twitter messaging, blogs, discussion groups, image and video sharing). One of the main benefits of publishing on media channels is discovery, influence, and navigation to your WISP promotional media.

Media Channel Types

Media channels can be owned (completely controlled), managed (moderated), or contributed (media sharing). one-way (broadcast), unicast (direct), two-way (social), or response (submission and monitoring).

Media Channel Setup

The setup of media channels starts by searching for and identifying available media channels that have category and keywords. Next, the account is created, name and URL selected, and the media channel configured for publishing and distribution.

Media Channel Management

Media channel management involves selecting the media message topics, content sources, and how often you should publish to the media channel. There may be multiple people who can publish to the media channel (staff, contractors, etc).

Media Channel Purpose

The purpose of media channels is to send or share messages to people or help them to discover your WISP service, influence their mindset about your WISP system or related products, and provide them a way to purchase your services or another followup activity.

Media Channel Objectives

Media channels are used to share messages with people at specific times to make them aware and influence their mindset about a product, service, or topic. The content (posts or media) published on media channels should be valuable to the audience, controllable by the sender, and measurable (sending and engagement if possible). Key objectives can be the types, content value, and number of media posts that are published. You may be able to measure or estimate the number of people who can see the media (reach) and how often the media posts are seen by the audience (frequency).

Media Channel Audience

Media channels may be used for their existing audience persona types (such as mature homeowners) or recipients may be chosen (targeted) from a media channel based on characteristics (such as age, location, income, etc).

Media Channel Reach

The number of people (audience size) that a media channel reaches is it's reach. Reach can be measured by the number of people who see your message (impressions.) In general, you get more influence (engagements and ability to remember) when people see your message multiple times (impression frequency).

Media Channel Content

The content that is published on a media channel should contain topic related information (within the channel scope). Media channel content may be sequenced to build awareness and change audience mindset.

Media Channel Frequency

Media channel frequency is the number of messages that are sent over time and how many times a recipient may see the same or a related ad. In general, it takes multiple impressions for people to become aware of your ad, remember it, and take some action.

Media Channel Types

Media channels are communication paths between the channel owner or manager and others who can read, use, and/or respond to the communication. Media channels can be owned, managed, or shared by promoters. Media channels can be one-way (web pages) or two-way (discussion groups). Media channels can be owned, managed, or contributed. You may own, manage, or contribute to media channels.

Owned Media Channels

You can setup and own multiple media channels which provide full control over the media publishing content and distribution times. Owned media channels include websites, blogs, email, and directly managed media services. When setting up media channels, it is typically good to use the key topic or search words in your media channel name.

Managed Media Channels

You can register and use media channels on other platforms such as Facebook, Linkedin, Pinterest, Youtube, and many others. While you own the profile, the content you publish on these media channels must conform to platform content and use rules. Even if you don't plan to publish on some networks or platforms, it is good for you to register your name on multiple platforms to ensure you own your name.

Shared Media Channels

You can join and share WISP related media with groups, Reddit, Quora, Facebook Groups, etc. When publishing and interacting on shared platforms, you must conform to platform rules, the media channel owner rules, and follow social rules from the media channel members. WISP service promotion on shared media platforms needs to be indirect. It is ok to share a broadband Internet topic tip. Direct promotion (offers and discounts) on shared platforms can get you banned from the group.

Broadcast Channels

You can send messages to groups of people (broadcast) to develop aware-ness. This can be an ad on a streaming radio channel or podcast, an email message broadcasted to Internet service users or any other media channel that can send messages to groups of people.

Broadcast advertising is called push (interruption) marketing because recip-ients are not expecting the messages. This means your ad needs to develop attention (stand out) and rapidly develop interest. Ads may be for branding (to develop awareness) or to motivate actions (product promotions).

Direct Marketing

You can send messages to specific people (direct marketing) based on their characteristics (profile targeting). Direct marketing allows for selecting message types or customizing messages.

Social Media

You can publish and respond to information on shared media channels such as social networks, groups, image boards, and others. Publishing on these channels must conform to the platform rules (no illegal items, hate mes-sages) and social rules (unwritten guidelines setup by the members). WISP marketing on social channels usually requires indirect promotion such as sample guides, trial user requests and other media that is helpful or is of interest to the members.

Response Channels

You can get or monitor communication from your audience on response channels. Response channels may be controlled (such as submission forms) or they can be passive (such as Google alerts) monitoring.

Media Channel List

A media channel list (such as a spreadsheet) identifies, organizes, and provides guidance on which channels are available, what content to be published, and how often it should be used.

Media Channel Name

Setup or join media channels that have names which relate to your WISP services and products. The title should be short (1-5 words) and be easy to remember.

Channel URL

The channel web address URL should include title or topic keywords if possible. It can be helpful to include word separators such as hyphens or underscores. This can help people to quickly see the topic and it can help search engines to topic focus.

Media Channel Description

Create a 1 to 5 sentence summary of the topics covered on the media channel and the range and depth of what will be shared on the channel. You may define the types of audiences that will benefit from the channel.

Media Channel Strategy

Define the strategy that you will use to start, grow, and maintain the flow of content on the media channel. This can include initial post topics and frequencies, member building campaigns, and volunteers for group management.

Media Post Frequency

Identify the types and number of messages you want to publish or make available on each media channel. For example, publishing 50+ blog posts over a 1 year period that includes helpful guides, lists of tools and other topics. This means you should publish 1 to 2 blog posts per week.

Media Channel Management

Each media channel needs to manage content creation, timing coordination, and publishing agents.

Media Post Content

Make a list of key media post topics, content sources, and the formats (short, long, images, etc) that will be published on the channel.

Publishing Rules

Setup media channel publishing rules and guidelines, media post creation and approval steps, and how to respond when the rules or messages require special handling such as highly negative feedback or hate messages.

Media Post Scheduling

Define the desired time periods between publishing on the channel (message frequency). You may start with a short time period between posts during media channel launch (such as every 1-3 days) and slow the publishing as the channel becomes established. You can setup and use media publishing scheduling tools such as Hootsuite.

Channel Monitoring

Setup channel monitoring (such as Google Analytics) and a reporting review process that allows you to determine which channels are working and how to improve them. Monitoring processes can be manual (such as viewing the number of company reviews on Facebook) or they can be automatically monitored using services such as Google Alerts. Setup time periods for review and setup a chart for achievements and results (such as the number of company reviews on Yelp).

Media Agents

Setup services or media agents who can create and publish media posts. Identify which media channels they will use and setup access for them to your media channel. Create assignments and measurable goals for your media agents such as the number of media posts published each month. Create guidelines and rules that your media agents can use.

Appendix 1 - WISP Glossary

The definitions in this glossary are common and popular terms related to Wireless Internet Service Provider (WISP) systems. To get more definitions, visit WISPDictionary.com

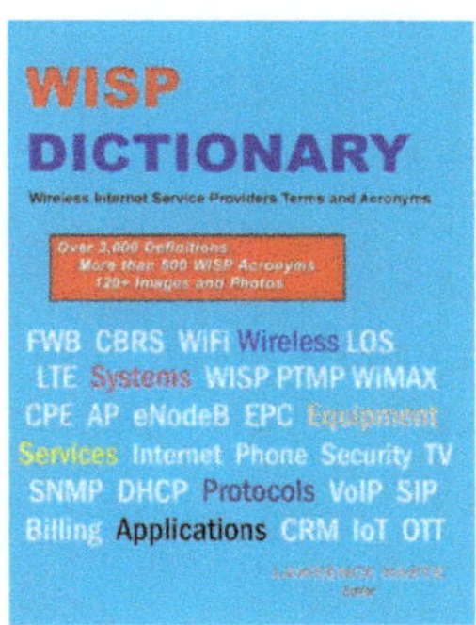

3G (Third Generation) - 3G is a mobile wireless technology that provides up to a maximum of 22 Mbps with typical data rates of 2 Mbps. This is good for text messaging (almost no delay), the sending of images in 1-3 seconds and for very low quality video.

4G (Fourth Generation) - 4G is a mobile wireless technology that provides for up to a maximum of 200+ Mbps with typical data rates of over 10 Mbps. This is good for photo sharing (almost no delay) and viewing of medium to high quality video with minimal buffering delays.

5G - (Fifth Generation) - 5G is a mobile wireless technology that provides for up to a maximum of 2+ Gbps (2,000 Mbps) with typical data rates of over 100 Mbps. This is very good for high quality video streaming and gaming with minimum delays.

6G - (Sixth Generation) - 6G is a mobile wireless technology in development that should be 10x faster than 5G (10 Gbps+). This is good for VR/AR, telemedicine, connected vehicles and other applications that require ultra high definition media with very short transmission delays. 6G networks are likely to combine multiple frequencies, systems and technologies using artificial intelligence (AI) to optimize connections and media flow controls.

Access Point (AP) - A radio communication location (point) that can connect users to a wireless system. An AP may be called a cell site, radio port or access node.

Average Revenue per User (ARPU) - the average money generated by all the customers of a business which is typically provided in a monthly amount. ARPU is an indicator of the operating performance of a business and may be compared to the ARPU for businesses with similar types of products and services.

Backhaul - The process of transferring communication signals over relatively long distances to a location for processing. This can be from Access Points to a point that can connect to the Internet.

Bandwidth - A term that defines the signal or data transmission rate capability. More bandwidth usually means faster file transfer rates and less delays in response time - great for rapid gaming control.

Billing and Support System (BSS) - are combinations of equipment and software that are used to allow companies to perform the service usage rating and product sales information. BSS systems can group this information for specific accounts for customers, producing invoices, creating reports for management, and recording (posting) payments made to customer accounts.

Broadband - A term that is commonly associated with high-speed data transfer connections where the data transmission is 1 Mbps or higher. Broadband Internet connections are not the same. Some can have much faster data transmission rates. A television channel needs 2 to 6 Mbps for standard TV and a 6 to 20 Mbps channel for HD television.

Buffering - The process of temporarily storing (buffering) transmitted data to create a reserve of media that can be used during gaps in transmission. Buffering can occur during the streaming of video (screen freeze) when the data transmission rates are not fast enough to provide for a live video streaming service.

Cable Modem (CM) - A communication device that converts signals on a cable TV system into data signals that can be connected to connected home devices such as smart TVs, computers, routers and other equipment.

Citizens Broadband Radio Service (CBRS) - A range of wireless frequencies (multiple channels) that can be shared between licensed and unlicensed users. For CBRS to work, when users (or their devices) request access to CBRS wireless channels, a spectrum allocation system (SAS) checks to see if there are any users already assigned to that channel. If they are, the CBRS system will coordinate access to other channels that are not currently being used.

Cloud Service Provider (CSP) - is a company that provides services through the Internet. A CSP owns or leases computer hardware and software systems that allow one or more users to access information services on or through Internet connections.

Communications Assistance for Law Enforcement (CALEA) - is a statute that was enacted by the U.S. congress in 1994 to define requirements of telephone service providers to provide wiretap capabilities to law enforcement agencies. To attach a listening device to a communication line, the law enforcement agency must have a surveillance order from a court of competent jurisdiction.

Customer Premises Equipment (CPE) - Communications equipment located on the customer's premises. This can include a radio transmitter, data switching router and other equipment. Fixed wireless CPE typically needs to be installed and precisely pointed to an Access Point. This usually requires a trained professional.

Customer Relationship Management (CRM) - is the process or system that coordinates information that is sent and received between companies and customers. CRM systems are used to schedule activities, allocate resources, and help control the sales activities within a company.

Data Cap - The digital transmission speed limit or a maximum amount of data that can be transferred over a period of time. While some systems claim to have unlimited data, companies may limit the maximum data transmission speed when users use a lot of data - such as from watching or streaming movies or videos.

Dead Spots - The areas within a wireless system (typically cellular or PCS) where, for one reason or another, signals do not have a sufficient level to provide an acceptable level of communications.

Dedicated Internet Access (DIA) - is a data connection that provides a constant speed connection in both directions.

Digital Media Adapter (DMA) - is a device or assembly that can convert digital media that is in one format (such as Internet data packets) into another format (such as HDMI or TV video signals).

Dedicated Service - A communication channel that is only accessible by one device to transfer user data. Dedicated services ensure that enough bandwidth is available at all times to transfer media such as streaming video or large file. Some systems such as cable TV networks share data channels between tens or even hundreds of users.

Digital Subscriber Line (DSL) - The transfer of digital information usually on a copper wire pair such as telephone lines. Although the transmitted information is in digital form, the transmission is a modulated analog signal that holds (represents) the digital information. The analog signal gradually gets distorted as distance increases which results in lower data transmission rates for connections that have longer wire connections - such as in rural areas.

Downlink - the radio connection from the system (access point) to the user device (customer premises equipment - CPE).

Download Speed - The rate at which information (digital bits) can be transferred from a system to a device over a time interval - typically in bits per second. Data transmission rates for Internet services are typically described in million (mega) bits per second (Mbps) or billion (Giga) bits per second (Gbps).

Dynamic Host Configuration Protocol (DHCP) - The process that assigns a temporary Internet Protocol (IP) address from a WiFi hub or router device, when devices want to connect to the Internet.

Evolved Packet Core (EPC) - Evolved packet cores are a portion of an 4G LTE network that is primarily designed to receive, process, and forward packets towards their destination. The use of an EPC provides rapid processing of packets, which increases data throughput while reducing packet delays.

Fiber to the Home (FTTH) - The transfer of digital information to a home through a clear strand of glass or plastic. Although the transmitted information is in digital form, the transmission is a modulated light signal that holds (represents) the digital information. Fiber data connections have extremely high data transmission rate capabilities. However, the cost to install and maintain fiber connections can be high - especially in long distance connections such as in rural areas.

Firewall - A device or software process that checks and blocks unwanted data communication connection types. A firewall can be set up to block access to certain types of services such as gaming or mature websites.

Fixed Wireless Broadband (FWB) - The use of wireless technology to provide voice, data, and video service to fixed locations using a high-speed Internet connection. The use of fixed locations instead of mobile service allows for focused and dedicated signals increasing bandwidth, capacity and reliability.

Gbps (Giga (billion) Bits per Second) - The measuring of data transmitted each second related to 1 billion (thousand million) bits. If your data transmission rate was 1 Gbps, you could transfer a movie DVD in less than 1 minute.

Hotspot Service - The ability of a wireless device (such as a smartphone) to retransmit its signal as a WiFi connection. When connected as a hotspot, the data transmission rate will be limited to the mobile device's connection speed and quality. If the mobile hotspot signal is low (1 or 2 bar levels), the data connection may be very slow.

Hybrid Fiber Coax (HFC) - is a communication network that combines wireless connections (typically for customer access) with a fiber interconnection connection system. Some customers in a HFC network may get fiber connections to their homes or businesses.

Internet of Things (IoT) - are devices that can communicate information about things. IoT things can range from very simple measurement data (temperature, pressure, etc) to processed information (objects identified by artificial intelligence - AI). IoT devices may communicate with existing systems (such as 4G, LoRA or other systems) or may use unlicensed frequency bands to directly communicate with other IoT devices.

IP Address (Internet Protocol Address) - A single connection identifier code that is added to each of the packets transferred between a user's device (such as a computer or streaming TV app) and other devices or services provided through the Internet. The IP address is usually provided when the computer or device first connects to the Internet (dynamically hosted) and remains associated with the device until it is disconnected or turned off.

Internet Service Provider (ISP) - A company that connects user devices such as computers, smart TVs, security video cameras to the Internet. An ISP purchases a high-speed link to the Internet and divides up the data transmission to allow many more users to connect to the Internet.

Internet Exchange Point (IXP) - is a physical location through which Internet infrastructure companies such as Wireless Internet Service Providers (WISPs), ISPs and content delivery networks (CDNs) connect with each other.

Last Mile - The final portion of a communication connection. To a home or business, this can be telephone wires (twisted pairs), coax (TV wire), fiberoptic or a wireless connection to a nearby access point.

Latency (Response Time) - The amount of time delay between the initiation of a data transmission request (such as pushing play on a smart TV remote) to when the request is received and processed. In general, systems that have very high data transmission rates (such as 100 Mbps) tend to have very small response delays (low latency). This can be very important for gaming and interactive video services.

Licensed Spectrum - is an authorization for a company to use radio spectrum (frequency bands) within according to the requirements of the license. These requirements may include a type of service (such as Internet service), channel types (single or multiple channels), and power levels within a specific geographic area (amount of signal strength allowed).

Lightly Licensed Spectrum - is an authorization for a company to use radio spectrum (frequency bands) provided when they pay a small fee which authorizes them to provide services according to the requirements of the license. This may be for exclusive use or for shared use.

Line of Sight (LOS) - A requirement of a direct unobstructed path between a radio device (such as a home antenna) and a wireless communication system (radio Access Point). Some low interference obstructions such as leaves on a tree may be ok as they only reduce the wireless signal a small amount.

Local Area Network (LAN) - A private data communication network that uses high-speed digital communications channels for the interconnection of computers, smart TVs, security cameras, and related equipment in a limited geographic area (such as a home or business). LANs can use wired cable (data cable) fiberoptic cables, coaxial (tv lines), twisted-pair cables, or wireless (WiFi) to transmit and receive data signals. A LAN can be connected to the Internet through a gateway or other data access device.

Long Term Evolution (LTE) - A wireless technology that is between 3G and 4G. It was created to provide higher data transmission rates (typically 5-10 Mbps) and more features than 3G while the full 4G technology was being developed.

Mega (million) Bits per Second (Mbps) - The measuring of data transmitted each second related to 1 million bits. If your data transmission rate was 1 Mbps, it would take you several hours to transfer a movie DVD. If your data transmission rate was 200 Mbps, it would take you less than 5 minutes to transfer a movie DVD.

Mobile Data - The transfer of digital information to a wireless device (such as a smartphone) through a mobile phone system. In general, wireless data transmission on mobile systems tends to be lower and less stable than fixed and focused wireless connections.

Multiple Input Multiple Output (MIMO) - is the combining or use of two or more radio or telecom transmission channels (signal travel paths) for a communication channel (such as an Internet connection). The ability to use and combine alternate transport links provides for higher data transmission rates (inverse multiplexing) and increased reliability (interference control).

Near Line of Sight (NLOS) - is a wireless communication system that does not require a direct clear path (can have some obstructions) between the transmitter and receiver. For example, some WISP radio connections are ok with having some tree leaves in the radio path.

Network Management System (NMS) - is a combination of equipment and software that is used to setup, control, monitor, and manage the operation of a communication network.

Over the Top Television (OTT) - is the delivery of digital television services over (on the top of) broadband Internet connections. The quality of television services may be controlled and/or guaranteed if the underlying broadband connections have enough bandwidth.

Peak Data Rate - The maximum amount of information (data) that can be transferred over a wireless connection during a short period of time.

Point of Presence (PoP) - is a physical location that allows an interexchange carrier (IXC) to connect to other companies such as an Internet service provider (ISP) or a local exchange company (LEC).

Power over Ethernet (PoE) - The adding of electrical power to a pair of wires in a data cable (such as category 5 wire used in home networks), so it can be used to power devices that are connected to it (such as the wireless CPE radio mounted on the wall or roof of a house).

Public Safety Access Point (PSAP) - are facilities that receive and process emergency calls. The PSAP usually receives the calling number identification information that can be used to determine the location of the caller. The PSAP operator will then initiate and/or route calls to assist with the emergency situation.

Quality of Experience (QoE) - is one or more measurement of the total communications experience or the entertainment satisfaction from the perspective of the end user. QoE measures may include service availability, audio and video fidelity, types of programming and the ability to use and the value of interactive services.

Quality of Service (QoS) - is one or more measurement of desired performance and priorities of a communications system. QoS measures may include service availability, maximum bit error rate (BER), minimum committed bit rate (CBR) and other measurements that are used to ensure quality communications service.

Roaming - The capability to use wireless services from another communication system. While roaming, features may not be available and additional usage fees may be charged.

Router - A device that directs (routes) content (data or digital media packets) from one path to another device or router that is connected to a network. Routers can communicate with each other to announce and discover new connection paths.

Satellite Internet - A system that connects user devices such as computers, tablets and other data devices to the Internet by wireless connections via satellites transmitters above the earth. Satellite systems that use satellites in fixed locations (Geosynchronous Equatorial Orbit - GEO Satellites) need high gain dish antennas to boost signals to reach them 22,300 miles/36,000 km above the earth and the transmission delays are over ½ second. Satellite systems that use satellites in locations close to the earth (Low Earth Orbit - LEO satellites) can use more traditional antennas to reach them about 300 to 400 miles/480 to 650 km above the earth and have less than 1/10th of a second delay time.

Shared Service - A communication channel that is accessible by multiple devices to transfer user data. Shared services coordinate or distribute bandwidth to multiple devices as needed. If the amount of available bandwidth (maximum data transmission rate) is not enough, users (devices) in a shared service will receive reduced data transmission rates which could result in video buffering (temporary delays) or slower file transfer.

Self Organizing Network (SON) - are communication networks that can automatically detect, setup, configure, manage, optimize and repair themselves. The use of SONs allows system owners to deploy and expand systems with limited amounts of technical skills.

Set Top Box (STB) - is an electronic device that adapts a communications medium to a format that is accessible by the end user. Set top boxes are commonly located in a customer's home to allow the reception of broadcast and/or streaming TV signals on a television or computer.

Simple Network Management Protocol (SNMP) - is a set of industry standardized commands and data messages that can be used to communicate management information between the network management stations (NMS) and the devices (radio access points, routers, etc.) in the network. By conforming to this protocol, equipment assemblies that are produced by different manufacturers can be managed by a single program.

Spectrum Allocation System (SAS) - is a database that maintains the usage activity of a group of radio channels (spectrum) which allows for multiple systems to share the same radio channels.

Spectrum Auction - is a procedure used by governments which control spectrum access rights (department of communication - DOC) to allow companies and people to bid for radio spectrum usage rights. The winning bidder typically gets the exclusive rights to use spectrum according to the usage rights (for voice, data, any) for periods of time.

Streaming TV - is the continuous transfer of motion picture information (television media) that is sent through a packet based communications network (the Internet).

Subscriber Acquisition Cost (SAC) - is the combined costs that are associated with marketing and adding of a customer to a system or service. Subscriber acquisition costs may include equipment subsidy, sales commissions and associated marketing costs.

Unlicensed Spectrum - is an authorization for a company to share the use of radio spectrum (frequency bands) according to the requirements of the regulations. These regulations typically include access control (processes on how to share), channel types (single or multiple channels), and maximum power levels.

Unlimited Data (actually limited) - A digital transmission service that does not have a limit of data that can be transferred over a period of time. While there is no limit, the maximum data transmission rate of the channel determines the maximum amount of data the user can get over a period of time.

Uplink - the radio connection from the user device (customer premises equipment - CPE) to the system (WISP access point).

Upload Speed - The rate at which information (digital bits) can be transferred from a device to a system over a time interval - typically in bits per second. In general, data upload speeds tend to be lower than download speeds, especially on mobile communication systems.

Virtual Private Network (VPN) - are private communication path(s) that transfer data or information through one or more data network which is setup between two or more points. VPN connections allow data to safely and privately pass over public networks (such as the Internet). The data traveling between two points is usually encrypted for privacy.

Voice over Internet Protocol (VoIP) - is a process of sending voice telephone signals over the Internet or other data network. If the telephone signal is in analog form (voice or fax) the signal is first converted to a digital form. Packet routing information is then added to the digital voice signal so it can be routed through the Internet or data network.

Worldwide Interoperability for Microwave Access (WiMAX) - WiMAX is a wireless communications standards-based technology that is used in Point-to-Point and Point-to-Multipoint wireless networks. WiMAX systems have been transitioning to 4G and 5G platforms.

Wireless Fidelity (WiFi) - A communication technology that allows computers and workstations to communicate with each by shared wireless signals (no radio license required). WiFi technology is described and standardized in the 802.11 wireless LAN system technical specifications. Different versions of the 802.11 specifications (e.g. 802.11a, 802.11b, and 802.11n) have different capabilities, such as higher data transmission rates.

WiFi Repeater - A wireless device that can receive, boost and retransmit WiFi signals. WiFi repeaters can help extend the range of WiFi routers. Because they only receive and amplify the signals, repeaters do not increase the transmission capacity of the WiFi system.

Wireless Internet Service Provider (WISP) - A WISP is a company that connects user devices, such as computers, smart TVs, and security video cameras, to the Internet by high-speed wireless connections. A WISP purchases a high-speed link to the Internet, sets up radio Access Points which allow many users to connect to the Internet. The home devices typically first connect to a home data network (such as WiFi or a router) which are connected to a CPE radio access device that links to the WISP radio Access Points.

Wireless Local Area Network (WLAN) - A wireless transmission system that allows devices to communicate with other devices and a wireless router that coordinates their data transfer (a WiFi router).

Appendix 2 - WISP Acronyms

3GPP - Third Generation Partnership Project
5G NR - Fifth Generation New Radio
AP - Access Point
API - Application Programming Interface
ARPU - Average Revenue per User
BNG - Broadband Network Gateway
BSS - Billing and Support System
CBRS - Citizens Broadband Radio Service
CM - Cable Modem
CPE - Customer Premises Equipment
CRM - Customer Relationship Management
CSP - Cloud Service Provider
DHCP - Dynamic Host Configuration Protocol
DIA - Dedicated Internet Access
DMA - Digital Media Adapter
DSL - Digital Subscriber Line
EPC - Evolved Packet Core
FTTH - Fiber to the Home
FWB - Fixed Wireless Broadband
Gbps -Giga (billion) Bits per Second
GIS - Government Information Service
GPS - Global Positioning System
HFC - Hybrid Fiber Coax
IoT - Internet of Things
IP - Internet Protocol
ISP - Internet Service Provider
IXP - Internet Exchange Point
LAN - Local Area Network
LEO - Low Earth Orbit
LOS - Line of Sight
LTE - Long Term Evolution
Mbps - Mega (million) Bits per Second
MIMO - Multiple Input Multiple Output

MRR - Monthly Recurring Revenue
NLOS - Near Line of Sight
NMS - Network Management System
OTT - Over the Top Television
PoE - Power over Ethernet
PoP - Point of Presence
PSAP - Public Safety Access Point
QoE - Quality of Experience
QoS - Quality of Service
SAS - Spectrum Allocation System
SAC - Subscriber Acquisition Cost
SNMP - Simple Network Management Protocol
SON - Self Organizing Network
STB - Set Top Box
VoIP - Voice over Internet Protocol
VPN - Virtual Private Network
WAN - Wide Area Network
WiMAX - Worldwide Interoperability for Microwave Access
WISP - Wireless Internet Service Provider
WLAN - Wireless Local Area Network
XP - Exchange Point